Apple

by Maryellen Gregoire

Consultant:
Adria F. Klein, Ph.D.
California State University, San Bernardino

capstone
classroom

Heinemann Raintree • Red Brick Learning
division of Capstone

Apple trees grow leaves.

Apple trees grow flowers.

Apple trees grow apples.

They grow lots of apples.

People pick apples from apple trees.

People eat apples.

In the fall, apple trees lose their leaves.